Ministry of Agriculture, Fisheries and Food

Calf Rearing

Reference Book **10**

London: Her Majesty's Stationery Office

First published 1930

Sixth edition 1984

ISBN 0 11 242643 3

Contents

	Page
Introduction	1
Pregnancy, Parturition and Early Life	2
Nutrition and management of the in-calf cow	2
Management during calving	2
Calving difficulties	3
Colostrum	4
The protective role of colostrum	4
Factors affecting absorption	5
Orphan calves	7
Feeding	9
General principles	9
Rumen development	9
Nutritional requirements of the calf	10
Bucket feeding	11
The basic system	11
Type of milk substitute	13
Weaning	16
Early weaning compound feed	17
Variations on the basic early weaning system	19
Environment and Housing	23
Climatic environment	23
Air temperature	23
Relative humidity	24
Ventilation	25
Light	28
Structural environment	28
Social environment	32
Choice of accommodation	32
Group or individual penning	32
Natural or fan ventilation	32
Ridged or mono-pitch roofed buildings	33
Converted buildings	33
Follow-on accommodation	34

Page

Husbandry and Disease Control 35
Cleaning and disinfection 35
Bought calves 36
Calf diseases 36
  Navel and joint-ill 36
  Scour or diarrhoea 37
  Salmonellosis 37
  Respiratory disorders 37
  Calf diphtheria 38
  Lead poisoning 38
  Lice 38
  Ringworm 38
Routine operations 38
  Dehorning 39
  Castration 40
  Removal of surplus teats 41
Identification 41

Appendices

I Basic housing requirements 42
II Production targets 43
III Budgeting pro-forma 44
IV MAFF publications 45
V Other publications 46
VI ADAS Experimental husbandry farms rearing calves 47
VII Conversion tables 48

Introduction

The foundations of success of many systems of beef production are laid in the calf rearing stage. It is essential that losses are minimal and that the husbandry system adopted allows the calf to express its growth potential without increasing cost unduly. This is particularly true with systems of beef production that require optimum growth rates throughout the animal's life. Similarly, with milk production, if replacement heifers are to calve down at two years of age an efficient calf rearing system is of paramount importance.

Over the years ADAS Experimental Husbandry Farms have studied methods of feeding and weaning calves, housed in a variety of buildings and despite the experimental nature of some of the rearing methods practised, losses have averaged only 3·5 per cent.

This reference book is based largely on the results of EHF work together with relevant information from other research centres. Where appropriate, results are quoted so that a system to suit particular circumstances can be built up and decisions taken on the basis of experimental evidence.

Pregnancy, Parturition and Early Life

Careful management and planning can improve calf survival and subsequent performance. Particular attention must be paid to:

- nutrition and management of the in-calf cow
- management during calving
- calving difficulties
- colostrum intake

Nutrition and management of the in-calf cow

If strong healthy calves are to be born then the nutrition of the in-calf cow or heifer is of great importance, particularly in the last three months of gestation when the foetus makes most growth. As a general rule cows should gain condition during this period and be fit (condition score 3) but not fat at calving. Gross overfeeding may lead to calving difficulties and underfeeding may have adverse effects on the cow and the foetus.

It is important that the vitamin and mineral status of the cow is maintained if problems such as abortion, abnormal, or weak calves are to be avoided. The main danger is during late winter if feeds low in vitamin A, such as poor hay, straw or unsupplemented cereals have been fed. At birth the calf has little or no reserves of the fat soluble vitamins such as A, D or E and must rely on colostrum and milk for adequate supplies. These will not be available if poor quality diets are fed during pregnancy. During periods of likely deficiency, such as the winter months, a vitamin supplement should be added to the dam's ration, or supplied by injection. A supplement that contains about 8 million i.u. vitamin A should be added per tonne of compound feed. The alternative is to inject the cow twice with 500 000 i.u. one month apart during late pregnancy.

Exercise of spring calving cows tied in cowsheds is important in maintaining their health prior to calving.

Management during calving

Calving should take place in a field or in a clean, dry draught-free calving box. This should have good artificial light, a clean water supply and ideally should be gated along one side for easy access. In addition facilities for lifting an animal may be needed during a difficult calving.

If, on examination, a normal presentation is indicated the cow should be allowed to calve without assistance. Patience is essential and providing calving progresses the temptation to assist when the feet appear must be resisted. Premature interference is likely to result in the calf becoming wedged in the partially open cervix and death of the cow and calf may ensue.

Should a malpresentation occur or progress in calving cease then experienced or veterinary assistance will be required. Excessive traction should not be used and any external force should coincide with the contractions of the cow. Cleanliness is essential and adequate supplies of antiseptic and lubricant should be readily available.

Immediately after birth care should be taken to ensure that the mouth and nostrils are clear so that the calf can breathe. It may be necessary in some circumstances to hold it up by its back legs to clear its lungs. In problem cases the tongue should be pulled forward and if the calf is not breathing, the chest walls should be compressed and relaxed by hand. If the dam is reluctant to lick the calf dry it may be encouraged to do so by sprinkling some bran or salt on the calf. If this fails the calf must be vigorously dried with straw which will stimulate movement and respiration.

The navel must be treated with an antiseptic such as tincture of iodine and this process should be repeated at least once in the first 24 hours after birth. Finally before encouraging the calf to suck make sure that the cow's udder is clean and that she is not suffering from mastitis.

Calving difficulties (dystokia and calf mortality)

Calving difficulties may result in a weakened calf or dam, the death of either or both, reduced milk yields and delayed conception. The risks may be reduced by careful choice of sire and by the correct feeding of the cow in late pregnancy.

Choice of sire

Bulls of breeds with a large mature size, produce bigger calves and more calving problems than do bulls from breeds with a small mature size. Data in Table 1 is based on 14 000 matings.

Table 1. Calving difficulties and calf death in beef herds by breed of sire (MLC)

Breed of sire	Difficult calvings %	Calvings requiring surgery %	Calf mortality %
Charolais	9·0	1·2	4·6
Simmental	8·9	1·2	4·4
South Devon	8·7	0·9	4·2
Limousin	7·4	0·7	3·8
Lincoln Red	6·7	0·3	3·2
Devon	6·4	0·3	3·6
Sussex	4·5	0·2	2·1
Hereford	4·0	0·3	2·0
Aberdeen-Angus	2·4	0·1	1·6

In general cows from larger breeds have more trouble calving, especially when bred pure or mated to bulls of breeds of even larger mature size. Dams from small dairy breeds seem to produce less dystokia from matings to bulls from breeds of large mature size than do Friesians or Holsteins similarly mated.

AGE OF DAM

The incidence of dystokia is greater with heifers than with second or subsequent calvers. Therefore, when cross breeding, bulls of breeds of large mature size should not be used on heifers.

An investigation by ADAS into the incidence of calving difficulties in some 3 000 Friesian heifers mated to Friesian, Hereford and Angus bulls showed that the weight of calves at birth and calving difficulties were reduced when Angus bulls were used.

Table 2. Sire of calf, weight of calf at birth and difficult calvings (ADAS)

Sire of calf	No of calvings	Av weight of calves (kg)	Difficult calvings (%)
Friesian	1 482	36·7	12
Hereford	997	37·7	13
Angus	636	33·4	7

The same survey showed that the percentage of calving difficulties and calf mortality increased in younger and older heifers compared with those calving at $2\frac{1}{4}$–$2\frac{3}{4}$ years of age. The results are shown in Table 3.

Table 3. Age at calving, calving difficulty and calf mortality in Friesian heifers (ADAS)

Age (Years)	No of calvings	Difficult calvings (%)	Calf mortality (%)
Under $1\frac{3}{4}$	33	27·3	18·2
$1\frac{3}{4}$–$2\frac{1}{4}$	1 561	11·9	12·2
$2\frac{1}{4}$–$2\frac{3}{4}$	943	9·3	8·3
$2\frac{3}{4}$–$3\frac{1}{4}$	451	13·3	13·7
Over $3\frac{1}{4}$	21	23·8	38·1

Bull calves are invariably larger than heifer calves and result in more calving difficulties.

Colostrum

The protective role of colostrum

Calves receive no antibody protection from their dams prior to birth and they can only obtain this through drinking colostrum. There is abundant evidence to demonstrate the importance of colostrum in combating disease in general, and scours and respiratory disorders in particular. It must be remembered however that colostrum only contains antibodies against those diseases to which the dam has been exposed.

The degree of protection from colostrum can be measured by a simple test on the blood of young calves and the level of protection in terms of antibodies present is expressed as Zinc Sulphate Turbidity (ZST) units. These should be interpreted as follows:

ZST Range	Protection
0– 5 units	Little or no protection
6–15 units	Poor protection
16–20 units	Marginal protection
21 and over	Satisfactory protection

As antibody levels rise so the percentage mortality and the proportion of calves requiring veterinary treatment falls, while at the same time daily liveweight gains improve, as a result of the reduction in disease.

Factors affecting absorption

The level of absorption and protection achieved can be greatly influenced by attention to four important factors:

1. THE TIME OF FIRST FEED

The efficiency of antibody absorption decreases gradually from birth so that calves should be fed as soon after birth as possible, preferably within the first four to six hours.

Table 4. Effect of time of colostrum feeding on antibody absorption by calves fed the same amount of colostrum. (Selman, IE (1969) Ph. D. Thesis, (Glasgow)

	Time colostrum fed (hours after birth)	Antibody levels in blood at 48 hours (ZST units) Mean
Group I	1	23·0
Group II	5	17·7
Group III	9	12·5

However, absorption at a reducing rate continues for about 24 hours and therefore, even if the important period has been missed, some degree of protection still can be obtained by late feeding.

2. THE TOTAL QUANTITY FED WITHIN THE FIRST 24 HOURS

The more colostrum a calf drinks within the first 24 hours the better the protection.

Colostrum is a mild purgative and feeding too much at one feed may cause a digestive scour. If the calf is left with its dam it can be expected to drink up to eight litres during the first 24 hours. Scour, if it occurs following the initial clearing of the digestive tract should be temporary and last only a few hours.

If the calf is removed from the cow two litres should be fed as soon after birth as possible. A bottle with a plastic teat or a teat-bucket may have to be used. This should be followed by at least two and preferably three additional feeds of the same amount in the first 24 hours. Thus the total amount fed will approximate to that which the calf would have sucked. It is important not to feed more than two litres at any one feed. Adequate colostrum intakes in the first 24 hours are essential to allow for differences in antibody concentrations in colostrum which varies widely from cow to cow.

3. THE PRESENCE OF THE DAM

For reasons which are not fully understood maximum absorption of antibodies at the same level of colostrum intake takes place when the calf is left in the presence of its dam. Calves should be left with their dams for at least 12 and preferably 24 hours after birth.

Table 5. The effect of the dam's presence on antibody absorption. (Selman *et al* 1971. *Res. Vet. Sci.,* **12,** 1)

	Average colostrum intake to 12 hours (kg)	Antibody concentration at 48 hrs (ZST units)
Non-mothered (kept apart from dams except when sucking)	3·4	18.3
Mothered (kept beside dams but muzzled except when sucking)	3·8	31·2

Many calves if left unattended with their dam will fail to suck within the vital first few hours, and so there are advantages from assisting suckling.

Table 6. The effect of assisted suckling on antibody uptake (Bridget's EHF)

Treatment at calving	Calves with above adequate antibodies (%)
Assisted	67
Left alone	40

Effective supervision requires patience because considerable time (up to 20 minutes) is required to suck to satiation. Suckling is more effective in ensuring good colostrum intake than other systems of feeding and experiments have shown that up to 80 per cent of suckled calves have satisfactory antibody levels compared with only 40 per cent of pail fed calves.

Table 7. Suckling versus pail feeding of colostrum (Smith *et al* 1967. *Vet. Rec.* **80,** 664)

Antibody level	Calves in each ZST group (%)	
	Suckled	Pail fed
Low	13	40
Marginal	7	20
Satisfactory	80	40

4. SEASON OF CALVING

No seasonal variation in the antibody level of colostrum is known to occur, but several investigations have shown lower concentrations of antibodies in calves' blood during the winter period. These are probably due to a failure to adopt the procedures advocated above, whereas calves born at grass are often left with their dams.

Table 8. Seasonal variation in antibody levels in bought calves (Drayton EHF)

	Percentage of calves in each group when born in:—			
	Sept	Nov	Feb	April
Low	8·4	11·7	6·7	15·0
Marginal	25·0	26·6	61·6	58·3
Satisfactory	66·6	61·7	31·7	26·7

Consequently rearers purchasing calves during the winter months should make allowance for the likelihood of a lower level of protection.

Orphan calves

If the dam's colostrum is not available, colostrum from other cows or milk from cows which have to be pre-milked can be fed fresh or can be stored, at −18° to −25°C, for up to 6 months.

- fresh colostrum should not be mixed with stale colostrum more than two days old.
- colostrum should not be mixed with water.
- colostrum should not be heated above 50°C as this destroys the antibodies and denatures whey protein.

If fresh or stored colostrum is not available then the following substitute may be fed:

1 whipped egg

0·3 litre water

0·6 litre whole milk

0·5 teaspoonful castor oil.

The above mixture should be fed three times per day for the first three to four days of life. An injection of a Vitamin A, D and E preparation should also be given. Antibiotics should only be given on the advice of a veterinary surgeon.

In practice a proportion of calves is likely to have low antibody levels. This will not necessarily have adverse effects if the calves do not come up against a severe disease challenge.

From the fifth day calves can be fed fresh milk, fresh or sour colostrum or a milk substitute.

Feeding

General principles

The young calf has a stomach in four parts, but only the abomasum or true stomach is functional. The reticulo-rumen and omasum do not develop until the young animal begins to eat dry feed. This may be as early as 7–14 days of age.

The action of sucking or drinking causes the reflex closure of the oesophageal groove and the food by-passes the reticulo-rumen and enters the abomasum where digestion begins. If a greedy calf drinks too rapidly liquid may spill from the over-taxed oesophageal groove into the rumen and this can lead to digestive upsets.

Efficient digestion and utilisation of nutrients require that food passes at a suitable rate through the alimentary tract which is equipped with appropriate enzymes and absorptive mechanisms. These allow the breakdown of complex food molecules and absorption of the resultant products. The passage into the lower gut of undigested or unabsorbed feed residues as a result of overfeeding will provide a medium suitable for undesirable bacterial growth which may lead to diarrhoea.

During the first few weeks of life when calves are being reared mainly on liquid feeds, the only major nutrients which can be efficiently utilised are casein or other high quality proteins, butterfat or homogenised vegetable or animal fats and the sugars lactose and glucose.

After milk is ingested by the calf, enzyme action causes clotting in the abomasum within a few minutes. During the first three to four hours after feeding, whey containing soluble sugars and nitrogenous compounds is released from the clots into the abomasum and passes down the alimentary tract. It is followed later by the remainder of the material containing protein and fat. Any impairment of the milk clotting mechanism is an important factor in predisposing calves to *E. coli* intestinal infections.

Rumen development

The development of the reticulo-rumen and omasum starts when the calf first takes solid food and is complete a few weeks after it is weaned. The longer a calf has access to a plentiful supply of milk the less will be its urge to supplement its diet with other feeds, and the slower will be rumen development.

Given a limited supply of milk or milk substitute, calves begin eating palatable dry feed at about seven days of age. These dry feeds pass into the rumen where bacteria and other micro-organisms become established and convert fibrous and starchy feeds into available forms of energy.

Once the rumen begins to function the incidence of digestive scour in calves given a balanced ration is negligible.

Encouraging calves to eat compound feeds rather than hay or other roughages aids the development of rumen papillae. These papillae increase the surface area of the rumen wall and thus the area through which nutrients can be absorbed.

Nutritional requirements of the calf

The maximum voluntary intake of dry matter depends upon the form of the feed. Most calves under 50 kg liveweight consume more dry matter per day as milk than as dry feed. In general calves eat compound feeds in preference to roughage, although high quality hay may be preferred to unpalatable or dusty compound feeds.

WATER

If a calf is receiving a large part of its feed as milk or milk substitute then the need for water is small. However, in an early weaning system the eating of dry food by a calf depends upon the animal drinking water right from the start. Young calves which are reluctant to drink water may be encouraged to do so by having it offered to them slightly warmed.

Fresh water may be supplied in buckets or in constant level water bowls. A guard rail prevents animals fouling their supply.

MINERALS AND VITAMINS

Proprietary compound feeds will normally contain adequate concentrations of minerals and vitamins. However in home mixed rations it is important that adequate quantities are included.

Mixes should be compounded to supply 14 g calcium, 9 g phosphorus and 0·7 g magnesium per day to each calf. In addition trace elements and vitamins A, D and E will be required.

ENERGY AND PROTEIN REQUIREMENTS

For practical purposes the total needs of the pre-ruminant calf are expressed in terms of litres of whole milk or reconstituted milk substitute, and the

Table 9. Daily requirements of a calf for metabolisable energy and digestible protein (calf gaining 0·5 kg/day)

Body weight (kg)	Metabolisable energy (MJ)	Digestible protein (g)
50	19	150
100	27	253

energy and protein requirements of the older calf in the form of metabolisable energy and digestable protein.

Bucket feeding

Most home reared and purchased calves are fed by bucket or modified teat arrangement.

A good target is for bull calves to gain 50–60 kg between birth and three months old. This means that home-bred calves of 36–40 kg at birth will weigh 90–100 kg at three months old, and that purchased calves of about 45–50 kg will weigh 100–110 kg.

The basic system

Home-bred calves should be allowed to suck their dams for the first 12–24 hours and then be bucket fed warm colostrum at least twice daily for the next three to four days. After that they should be treated similarly to bought-in calves.

The source of supply of bought-in calves, and the type of calf purchased has a marked effect on the success of a calf rearing enterprise. Ideally calves should be purchased direct from the farm of birth to minimise stress in transit and reduce the risk of infection. However, the majority of purchased calves change hands through the open market, where they come into contact with calves from other sources, and often travel long distances to the rearing farm.

On arrival all calves should be checked carefully for signs of scour, rupture, wet navels, colds and other signs of ill-health, and then put to rest in a warm, dry, well littered calf pen. This is one of the few occasions when supplementary heat may be justified.

Purchased calves often appear hungry on arrival, but, due to stress factors their ability to digest milk may be impaired. Thus their intake of milk should be restricted until they have had at least six hours to settle down. In practice calves arriving in the morning need not be fed until the evening, and calves arriving in the afternoon/evening need not be fed until the following morning.

Treatments such as weighing and injecting are best left until the day after arrival. An injection of vitamins A and D is often worthwhile, especially with calves purchased in January, February and March as their dams may have been on vitamin deficient diets in late winter.

There is considerable divergence of opinion as to whether a glucose solution should be given to calves on arrival. If the transit journey has been short and the calves have recently been fed, additional carbohydrate (in the form of glucose) may cause scour. On the other hand if the calves have been without food for a considerable time then feeding glucose (50 g/1 water) as a readily available source of energy can be recommended. If any sign of scour is evident it is wiser to feed water alone or water plus two teaspoonsful of salt per litre.

Once newly purchased calves have settled down the feeding should consist of small feeds, increasing as calves become acclimatised to their new surroundings. Hay, water and a palatable early weaning compound feed should be on offer after the first day. The appetite for solid food is limited during the initial rearing period, and consequently only small amounts of hay and early weaning compounds should be provided, and renewed daily. Quantities should be increased as the appetites of the calves increase.

The basic and probably most successful system of bucket feeding is to feed twice daily. The schedule for home-bred calves from day 5 or for bought-in calves from the first full day on the farm is shown below. The quantities refer to a typical milk substitute mixed at 100 g powder per litre of water.

Day 1	0·75 litres	twice daily
Days 2–3	1 litre	twice daily
Days 4–5	1·5 litres	twice daily
Days 6–7	2 litres	twice daily
Day 8 onwards	2·25 litres	twice daily – full quantity of feed
Week 5	weaning takes place	

All calves should be treated as individuals. In the early stages shy feeders may require their daily allocation in three feeds, and the increase in feeding over the first week can be varied according to calf health and performance. Variations should be made in the quantity rather than strength of milk substitute offered, and a standard concentration of 100 g milk substitute per litre of water is recommended. (This supplies 450 g milk substitute per day when on full feed.) Possible variations when rearing on milk substitutes are:

1. Concentration of substitute
2. Temperature of milk substitute

CONCENTRATION OF MILK SUBSTITUTE

Feed according to the manufacturer's instructions. Stronger mixtures have been shown to produce no benefit.

Table 10. Effect of concentration of milk substitute on performance of beef x dairy calves fed twice daily (Liscombe EHF)

	Concentration (g/1)	
	100	200
Initial weight (kg)	45·3	48·6
Days to weaning*	49·0	42·0
Weaning weight (kg)	68·2	68·0
Gain/day to weaning (kg)	0·47	0·46
Post weaning gain to 12 weeks (kg/day)	0·86	0·86
Milk substitute fed (kg)	19·8	27·4
Compound feed fed to weaning (kg)	35·6	18·8

*Weaning based on live weight of 68 kg

Increasing the concentration of the substitute reduced the time to weaning but increased the cost per kg liveweight gain.

In general terms the stronger the solution the greater are the chances of scouring, and the weaker the solution the more bedding is required. However, if once daily feeding is practised (see page 20) the substitute must be mixed at an appropriate rate to ensure that the calf receives 450 g of milk substitute/day e.g. 3 litres @ 150 grams per litre.

TEMPERATURE OF MILK SUBSTITUTE

It is generally recommended that milk or milk substitute should be fed at a temperature of 35–38°C, and this will ensure maximum intake. Alternatively milk may be fed at ambient temperature. Constancy of temperature from feed to feed is more important than actual temperature.

Trials with cold milk substitute fed once or twice daily have given disappointing results during winter months in northern areas – calves have had to be coaxed to drink their milk when temperatures fell below about 5°C and some refused to drink completely. Cold milk feeding has succeeded in southern areas, especially in the warmer months, as shown in Table 11.

Table 11. The performance of heifer calves fed twice daily on warm or cold milk substitute (Bridgets EHF)

	0·45 kg milk sub in 3·4 litres/day	
	Warm	Cold
Birth weight (kg)	36·7	37·7
Weaning weight (kg)	49·9	49·0
12 week weight (kg)	86·6	84·8
Liveweight gain/head/day (kg)	0.59	0.56
Days to weaning	33	32
Compound to weaning (kg)	10·0	11·3

From trials that have been carried out it is clear that the temperature of milk substitute can be reduced considerably without causing trouble with scouring. Trouble has been encountered, however, when milk has been fed much warmer than blood heat.

Cold milk substitute feeding has advantages in big units and enables a more constant feeding temperature to be maintained.

Type of milk substitute

CONVENTIONAL POWDERS

Low fat (1–5 per cent fat in the dry matter) and high fat (15–20 per cent fat in the dry matter) powders are available.

Liveweight gains have been higher on high fat diets and have been associated with a reduced incidence of nutritional scours. However, because of their markedly lower cost some saving may be made by feeding low fat substitutes where housing and management are of a high order.

Table 12. Mean performance of dairy heifer calves fed 0·48 kg low or high fat milk substitute once daily (Leaver, JD and Yarrow, NH *Anim. Prod.* **14** 155–159)

	Low fat (4%)	High fat (17%)
Initial weight (kg)	42·8	42·2
Liveweight gain (kg/day)		
– to weaning	0·29	0·36
– to 8 weeks	0·51	0·56
Compound feed intake (kg)		
– to weaning	6·5	7·4
– to 8 weeks	50·2	53·7
Incidence of scours (calf days)	31	13

Powders in which casein has been partially replaced

With the increasing cost of milk protein (casein) alternative sources have been sought for use in milk substitutes. Partially replaced casein milk substitutes are manufactured in such a way that part of the dried skim milk solids, which form about 80 per cent of conventional milk substitutes, are replaced by up to 40 per cent animal or vegetable protein, usually of the fish meal or soya bean meal type. These substitutes for casein have to be carefully selected because of the limited range of enzymes present in the young calf's digestive tract.

Because the casein substitutes remain in the abomasum for a shorter period than conventional milk proteins there is the possibility of an increased incidence of scour, and this should be balanced against any cash economies that may be made. Farm trials have shown that when fed once daily at the same strength as conventional milk substitutes, (200 grams per litre) compound feed intake was lower and liveweight performance to weaning was reduced. However, when fed at 300 grams per litre performance was similar to that achieved with conventional powders.

Non-milk protein replacers tend to fall out of suspension and need agitating over the feeding period. This could be a problem with a large batch of calves.

'Zero milk' replacers

More recently a new generation of milk replacers termed 'Zero milks' has been introduced. These are usually based on non-milk proteins. Zero milk replacers do not attract EC subsidies and manufacturers must make use of sources of energy and protein which have lower unit costs than those in skimmed milk powder. They are based on a selection of whey products, soya protein products, fish protein hydrolysates and partially hydrolysed cereal starch.

There is little evidence in this country of the advantages and disadvantages of this type of milk substitute but some products have been extensively tested in Europe. Current trials at Bridgets EHF with a 'zero milk' blended from whey, soya, casein, animal fats and vegetable oils, and fed warm on a twice daily feeding system have resulted in satisfactory performance and a saving in feed cost.

ACIDIFIED MILK REPLACERS

These were developed to prolong the keeping quality of the reconstituted milk so making these powders suitable for cold *ad libitum* feeding systems. It is claimed that acidified milk replacers are more convenient and labour saving than conventional milk substitutes and that *ad libitum* feeding enables a calf to drink little and often so reducing the risk of scours and improving growth rates.

Acidified milk replacers are available in two basic types, medium acid (pH 5·6–5·9) and high acid (pH 4·2–4·4). Composition varies between brands with a considerable range of fat and protein contents. The medium acid powders are normally made from dried skim milk plus added fats, or dried whey with additional fats and milk proteins. High acid milk powders however have to be based on whey with non milk proteins such as fishmeal or soya bean meal because casein (milk protein) coagulates at the higher levels of acidity used. In both types the acidity is increased by the inclusion of an organic acid which acts as a preservative. The reconstituted milks remain fresh for two to four days after mixing depending on the degree of acidity and air temperature.

It is generally recommended that acidified milks should be fed cold *ad libitum* to groups of calves with one teat to every four or five animals. Alternatively calves can be individually penned and fed individually or share a bulk supply. For group feeding large rust proof containers (plastic barrels) fitted with tubes and teats are required. The container should hold three days' supply which at peak intake could be 12 litres/calf/day. Such an arrangement is easy to set up and cheap to install.

Intakes of milk substitute average some 30 kg/calf which is twice that consumed by calves on restricted feeding systems. As a result of high liquid intakes calves are reluctant to eat solid food, and although growth rates pre-weaning are often better than when restricted quantities are fed, it is difficult to maintain this improved growth rate after weaning. As a result calves are little heavier at 12 weeks than those reared on restricted milk feeding systems.

High liquid intakes inevitably result in more urine being produced and good drainage and adequate bedding is essential. Weaning calves on *ad libitum* feeding system is best done by restricting access to the milk supply by initially removing the teat during the night and then for increasing periods during the day.

Palatability problems have occurred particularly with high acid milk powder and purchased calves have been slow to adjust to the feeding system, particularly in cold weather. Under frosty conditions intake is reduced considerably and may be inadequate to sustain the required rate of liveweight gain.

Six ADAS EHFs carried out a co-ordinated trial to compare conventional, medium acid and high acid milk substitutes. These were fed to home bred

Table 13. Calf performance and feed consumption on acidified milk replacers. Friesian bull calves (Drayton, Gleadthorpe and Liscombe EHFs)

	Conventional powder Bucket warm 2×day	Medium acid powder *Ad lib* teat	High acid powder *Ad lib* teat
Liveweight gain 0–10 weeks (kg)	0·59	0·61	0·62
Quantity of milk powder (kg)	17	28	30
Early weaning compound feed (kg)	94	78	83

and purchased calves from buckets or through teats. On some farms the calves were housed in individual pens and on others they were grouped with one teat to four or five calves. On all farms the calves had free access to water, a palatable early weaning mixture and good hay. Weaning took place by gradually diluting the strength of the milk substitute. Calves on *ad libitum* fed acid milk substitutes consumed more milk but less compound feed than traditionally reared calves. Liveweight gains were similar but the feed cost per kilogram of liveweight gain was lowest on the traditional system.

Weaning

The sooner the young calf is eating and digesting dry food the lower the incidence of scour, the easier the feeding routine and the cheaper the cost of rearing. It is therefore important to wean the young calf as soon as possible, and one of three criteria can be used:

- compound feed intake
- live weight
- age

1. ACCORDING TO COMPOUND FEED INTAKE

This system is most suitable for individually penned calves. Such calves must be eating sufficient dry feed to prevent undue liveweight check after weaning. EHF results show that 450 g/day is adequate for home bred calves and that there is no advantage in going as high as 900 g/day. Experience has shown that a safe compromise for bought-in calves is 750 g/day.

2. ACCORDING TO WEIGHT

Purchased calves for intensive or semi-intensive beef production can be weaned successfully when:

- they have gained 14 kg since arrival
- they weigh a minimum of 60 kg (although Charolais and other large framed crosses may need to be taken to heavier weights)

3. ACCORDING TO AGE

This system is suitable for group reared calves, and avoids the need for weighing.

Under good conditions purchased calves can be weaned at four weeks with a reduction in rearing costs, although liveweight gains to 12 weeks can be improved by delaying weaning to five weeks as shown below.

Table 14. Effect of age of weaning on calf performance (Boxworth EHF)

	Friesian		Hereford × Friesian	
	4 weeks	5 weeks	4 weeks	5 weeks
Liveweight gain (kg/day)				
- birth to weaning	0·30	0·34	0·30	0·36
- birth to 8 weeks	0·37	0·42	0·40	0·43
- birth to 12 weeks	0·53	0·59	0·50	0·54

Whatever criterion is chosen for weaning it is generally better to wean abruptly. Calves weaned gradually do not settle to eat dry food properly if they are constantly waiting for their liquid feeds.

Care must be taken to ensure that compound feeds are kept fresh. Group fed calves should be offered compound feed twice a day to appetite to give slow eating or shy calves a greater chance of having their full ration.

After weaning, early weaning mixtures or pellets up to a maximum of 2·25 kg/head/day should be offered, with good quality hay or silage fed to appetite.

Early weaning compound feed

An early weaning compound feed should be highly palatable, have a metabolisable energy (ME) value of not less than 12·6 MJ/kgDM and contain 18 per cent protein (CP) as fed. An example of an early weaning mix for calves up to three months old is shown below. It has an ME value of 13·0 MJ/kgDM and contains 18 per cent CP as fed.

	kg
Barley (rolled or coarsely ground)	575
Flaked maize	150
Fishmeal	50
Extracted soya bean meal	150
Molassed meal/molasses	50
Mineral/vitamin supplement*	25

*to provide 4 000 i.u. vitamin A and 1 000 i.u. vitamin D per kg final mix.

A lower crude protein content reduces cost but has an adverse effect on performance.

The use of urea to supply a proportion of the crude protein is not recommended for calves below 14 weeks of age.

Table 15. Effect of crude protein content of calf compound feed on calf performance (Drayton EHF)

	Crude protein level (%)		
	14	16	18
Daily liveweight gain (kg)			
0–5 weeks	0·48	0·53	0·57
0–12 weeks	0·55	0·62	0·68
Feed consumption to 12 weeks (kg)			
Compound feed	130	143	147
Hay	6·6	6·0	5·9

A simple mixture based on rolled moist barley can be effective as an early weaning feed, but mixtures based on dry stored ground barley may be dusty and unpalatable unless an adequate amount of molasses or molassed meal is included in the formulation.

Although compound feeds composed of a number of ingredients are generally preferred, a range of simple rations tested at Gleadthorpe EHF have given good results.

Table 16. Early weaner compound feeds based on moist barley (% composition) used at Gleadthorpe EHF

	Barley: Fishmeal	Barley: Linseed/ Soya	Barley: Beans
Rolled barley	88·4	77·0	48·1
Milled field beans	–	–	48·1
White fish meal	9·8	–	–
Linseed meal	–	9·6	–
Soya bean meal	–	9·6	–
Sterilised bone flour	–	1·8	1·8
Limestone flour	1·2	1·2	1·2
Common salt	0·3	0·6	0·6
Trace mineral + vitamin A and D supplement	0·2	0·2	0·2
Calculated crude protein per cent as fed	17·5	16·2	19·2

There is little evidence that pelleting increases the intake of compound feeds by calves. Pelleting does however, improve the uniformity of consumption of mixtures in which some ingredients might otherwise separate out, but very hard pellets have been shown to reduce intake.

Work at Boxworth EHF evaluated high quality dried grass as an early weaning feed for calves. Although calf performance on a diet of all dried grass was reduced, a 50:50 mixture of dried grass and early weaner compound feed gave good results. The dried grass used was R V P ryegrass cut in mid-April having a D value of 74 and a crude protein content of 22·9 per cent. The results are shown in Table 17.

Table 17. Performance of calves fed chopped dried grass as a weaning ration

	Compound feed only	Dried grass only	50/50 mixture
Live weight (kg)			
at birth	38·6	40·0	38·4
at 12 weeks	85·4	79·9	81·5
Live weight gain (kg/day)			
birth - 12 weeks	0·55	0·47	0·51
Feed consumption (kg)			
birth - 12 weeks	107	99	108

Variations on the basic early weaning system

The basic system described above can be modified, in order to reduce costs or save labour, by:

- feeding whole milk
- feeding surplus colostrum
- once daily feeding
- unrestricted cold milk feeding
- automatic-machine feeding
- rearing at pasture

Often such systems require a higher level of stockmanship.

FEEDING WHOLE MILK

This is at present more expensive than feeding milk substitute, but when fed at the rate of 2·5–3 litres/head/day on an early weaning system, it results in good intakes of early weaning compound feed and good calf performance.

It has been calculated that 1 litre of whole milk is equivalent to 170 g high fat milk substitute powder, (Leaver J D, and Yarrow N H, *Anim. Prod.* 1972 **15** : 315–318) and this enables the economics of whole milk feeding to be calculated.

FEEDING SURPLUS COLOSTRUM

During the four days after calving a Friesian cow produces some 45–64 litres colostrum which should not be offered for sale. This is almost enough to rear an early weaned calf. Thus on a dairy farm, selling bull calves and rearing only heifer replacements enough surplus colostrum is available to replace totally the use of milk or milk substitutes. This work, pioneered at Bridgets EHF and tested on other EHFs has shown that heifers can be reared successfully on various systems incorporating colostrum:

	Weight at 12 weeks (kg)
Milk substitute throughout	84
Colostrum for 14 days followed by milk sub.	85
Colostrum throughout	86
Colostrum and milk substitute alternated (5 changes to weaning)	83

Calves were normally fed cold stored colostrum after the third day at the rate of 2·25 litres per day. Increased quantities gave only slight improvements in performance, and feeding 4·5 litres per day led to scouring.

Good hay, early weaning compound feed and clean water should be available from the start.

Surplus colostrum can be stored in a 'stock-pot' covered to keep out flies. Under normal weather conditions the acidity rises rapidly over the first 12 days, the product taking on the consistency of yoghurt. The acidity acts as a preservative, and stored colostrum is still acceptable to calves after 75 days.

In cold conditions, especially in the North, stored colostrum may need warming (to approximately 20°C) to assist the rise in acidity and may also need warming before feeding to ensure an adequate intake.

ONCE DAILY FEEDING

The aim of this system is to allow the peak labour demand for calf rearing to be allocated to a period of the day when work schedules are relatively slack. It also reduces the time involved in mixing and bucket washing, thus ensuring that highly skilled labour can spend more time with the calves.

This system requires a high standard of management. Calves are fed twice daily for a week building up to 450 g of powder per day, and then changed onto once daily feeding over a period of about three days. High quality milk substitutes must be used (EHF work was based on high fat substitutes) and calves should be fed at the same time each day. Clean water must be available at all times.

Calves fed once daily on warm milk substitute have performed as well as those fed twice daily provided an equal amount of milk powder is fed (i.e. the concentration of the made-up milk substitute must be increased to allow for the reduced liquid intake).

Table 18. Frequency of feeding on the performance of beef type calves fed 0·45 kg milk substitute powder/day (Randall, E M and Swannack, K P 1975. *Expt. Husb,* **28,** 44–52)

	Once daily	Twice daily
Arrival weight (kg)	45·6	45·1
Weaning weight (kg)	56·2	54·9
12 week weight (kg)	100·2	100·1
Liveweight gain/head/day		
arrival to weaning (kg)	0·59	0·49
arrival to 12 weeks (kg)	0·65	0·66

Because of the reduced frequency with which calves are seen they must be inspected more closely at feeding time to identify any animals that are off colour.

UNRESTRICTED COLD MILK SUBSTITUTE FEEDING

The aim here is to avoid the labour of hand feeding and the expense of automatic calf feeding machines. The system involves normal feeding for 8–10 days over which period calves are fed twice daily, the quantity of powder fed is increased to 450 g/day and the temperature is reduced so that by day 8 the calves are receiving cold milk.

Thereafter unrestricted milk is available via teats connected to churns or dustbins of milk substitute. Initially it is fed at a strength of 50 g powder per litre but this is stepped up over a two day period to 100 g powder per litre of water.

This system has worked well at Liscombe EHF and has matched the performance achieved on twice daily feeding of warm substitute. Milk substitute consumption with light calves has been high – up to 27 kg with calves initially weighing 43 kg (compared with 15 kg for calves weighing 48 kg).

This system requires a high degree of stockmanship to distinguish healthy calves that are 'loose' because of the system of feeding, and those that are scouring. Experience suggests that it is not suitable for use in the winter months, but that it works well at grass in the summer.

AUTOMATIC MACHINE FEEDING

Automatic feeding is a refinement of the above system which allows unrestricted access to warm milk substitute prepared automatically by a machine.

There are savings of up to 30 per cent in labour and the constant supply of freshly mixed milk substitute most closely resembles the normal relationship between calf and cow. Liquid intakes are higher so that not only are larger quantities of powder fed per calf but more bedding is required to keep calves dry. A high level of stockmanship is required, to spot calves that are not thriving. The machines have high capital costs and depreciate rapidly.

Trials have indicated good performance on these machines, and some results are summarised in Table 19.

Table 19. The performance of beef type calves on automatic feeding (Gleadthorpe and Liscombe EHFs)

	Automatic machine (Nursette)	Twice daily
Arrival weight (kg)	44·5	47·7
Wean weight (kg)	59·2	62·6
12 week weight (kg)	106	109
Liveweight gain/head/day		
Arrival to weaning (kg)	0·45	0·50
Milk substitute used (kg)	17·7	11·0
Compound feed used (kg)	129·0	148·0

Milk substitute intakes were increased and there were marked reductions in labour requirements.

Care must be taken in introducing calves to the system to ensure that all calves drink at least twice daily. The milk substitute should be mixed at 50 g per litre initially, rising to 100 g per litre over a 15 day period.

Feed nipples, water bowls and feed troughs should all be close together and after weaning a constant supply of fresh water should be provided if compound feed consumption is to be maintained.

REARING AT PASTURE

It is a traditional belief among farmers that young calves should not be turned out to graze in the year in which they are born. Spring born calves are often housed during the whole of their first summer and are not turned out until at least 12 months old. The main reason for this is the fear that the calves will become infected with husk and gastro-intestinal worms. The young calf can digest grass efficiently from about three weeks of age and it is therefore possible to rear it on pasture from an early age provided the grass is supplemented with compound feed until adequate quantities of herbage are eaten.

To avoid problems with worms best results are obtained if the rearing pasture is maiden seeds or a sward which did not carry calves or young cattle in the preceding year.

Trials have shown that calves can be reared successfully at grass using a variety of systems.

On arrival calves should be penned close to the feeding point, and hay, water and compound feed offered. As they settle down they can be given more land.

After weaning compound feeding should be continued at 2–3 kg daily, until calves are 12 weeks old. Then, if the grazing is of good enough quality, the level of compound feeding can gradually be reduced.

Protection should be provided for young calves as much against extreme heat as against cold. This, however, need not be elaborate or expensive.

Environment and Housing

Accommodation for calves need not be elaborate. The main requirements are a dry, draught free bed and a constant supply of fresh air. The design of the building should provide good working conditions and allow ease of supervision and management. Capital costs should be kept at a minimum consistent with efficiency and construction should permit thorough cleaning and disinfection.

In addition to providing for the basic environmental requirements for the optimum performance account must be taken of Leaflet 701 *Codes of recommendations for the welfare of livestock. Cattle* (1983).

The environmental requirements of calves include all those aspects of the calves' surroundings which affect its well being and performance. It includes climatic, structural and social factors and their interaction with nutrition and health.

Climatic environment

Air temperature

The calf, like other warm blooded creatures, maintains a relatively constant body temperature by balancing its heat production and heat loss. Heat production is primarily governed by the level of feeding and heat loss by coat insulation, air temperature, air speed and the nature of the floor and other surfaces with which the calf comes into contact while lying down. Air temperature, as such, is not normally of major importance to the healthy well fed calf given a dry bed and subjected to low air speeds. The air temperature below which the calf shivers or engages in some physical activity to keep warm is termed the lower critical temperature (LCT) and this is directly affected by all the factors which influence heat production and heat loss. Critical in this context simply indicates the temperature at which the calf begins to use energy (feed) to keep warm and does not imply that the calf is necessarily endangered in any way.

The main factors affecting heat loss which are of practical importance are floor type and air speed at calf level. The effect of floor type on the LCT of calves has been calculated by Mitchell (Calf Housing Handbook 1976) who showed that a dry straw bed had a marked effect on the lower critical temperature of calves, i.e: 8°C compared with 17°C for 50 kg calves lying on dry concrete.

Mitchell also showed the effect of nutrition on the lower critical temperature of the calf. The higher the energy intake; the greater the heat production; the lower the critical temperature.

As air speed at calf height increases the insulation value of the coat is reduced and heat loss increases. The effect this has on the lower critical temperature has been calculated by Webster (Optimal Housing Criteria for Ruminants 1981) and is shown in Table 20.

Table 20. Effect of air speed, calf weight and feed level on lower critical temperature (Webster 1981)

Type of calf	Lower critical temperature °C at air speeds of:	
	0·2 m/s	2·0 m/s
New born (35 kg)	+ 9	+17
One month old (50 kg)	0	+ 9
Veal calf (100 kg)	−14	− 1

The table emphasises the importance of ensuring that calves have access to conditions of low air speed and that draughts are eliminated from calf accommodation. Air speed close to the calves should never exceed 0·25 m/sec in winter. This can be achieved by having protected (usually perforated or slatted) inlets above calf level and the use of partial covers over the rear portion of the pens to prevent down draughts.

Experimental evidence and practical experience support the view that in draught free conditions air temperature, by itself, is not of major importance to the healthy, dry bedded well-fed calf. The effect of air temperature on calf performance was studied at Drayton and High Mowthorpe Experimental Husbandry Farms over a three year period from 1968–1970. Calves were reared throughout the year at ambient temperature or under regimes where the temperature was controlled at 7°C or 18°C. Calf performance was unaffected by treatment.

Records from EHFs show that air temperatures in naturally ventilated buildings closely follow external temperatures. The mean temperature rise being 1°–2°C with a further 2°–3°C rise occurring at calf height under partially covered pens. In practice supplementary heating is only likely to benefit dis-advantaged (starved, stressed or sick) calves or new born calves under extremely cold conditions.

Relative Humidity

The direct effects of relative humidity (RH) are twofold. At high ambient temperatures high humidity restricts the loss of body heat by evaporative cooling, and at lower levels of relative humidity evaporation takes place freely giving rise to a rapid increase in heat loss. At low ambient air temperatures high humidity increases heat loss from the body as humid air is a poorer insulator than dry air.

The indirect effects of high RH are likely to be the most important practical consideration. High RH at low temperatures invariably results in condensation on cold surfaces. This has a deleterious effect on the building structure and results in damp bedding and unhealthy conditions under which the incidence and severity of pneumonia are likely to increase.

Humidity within buildings is influenced by stocking density, feeding system, temperature, ventilation rate and drainage. Calf rearing systems involving high liquid intakes, such as *ad libitum* milk feeding, produce copious quantities of liquid. High temperatures increase evaporative heat loss and thus add to the moisture load. Control is effected by minimising the surface area of liquid and the time of exposure to the atmosphere. Floor slopes of a minimum of 1 in 20 under the animals and a drainage system which conducts the liquid rapidly out of the buildings are required. The preparation of liquid diets and washing of utensils should be carried out in a separate air space and the use of large quantities of liquid for swilling down passages avoided when buildings are stocked. Adequate ventilation rates assist humidity control by removing moisture as it is produced and also by minimising temperature rise. High stocking densities clearly increase the rate of moisture production per unit of space.

In practice the aim must be to keep relative humidity as close as possible to outside RH by good ventilation and drainage.

Ventilation

The primary purpose of the ventilation system is to remove the water vapour, heat and airborne micro-organisms given off by the calves together with dust and noxious gases produced within the building. If the ventilation system achieves all those objectives it will also provide an adequate supply of fresh air. All of this must be accomplished without subjecting the calves to draughts.

In most cases it is possible to arrange adequate ventilation by natural means. Wind pressure is the dominating force in naturally ventilated housing for most of the time but in order to ensure satisfactory ventilation in the absence of wind it is necessary to design the system to work on the stack effect. This depends on the heat given off by the stock rising to the highest point in the building, escaping and being replaced by colder air coming in from outside at a lower level. The essential features of good natural ventilation design for calves therefore are:

- An adequate sized outlet at the highest point of the building.
- Evenly distributed inlets below the level of the outlet but above the level of stock.
- Inlet areas larger than the outlet.
- Inlets designed to minimise the effect of high wind speeds and avoid draughts impinging on the stock.

Natural Ventilation

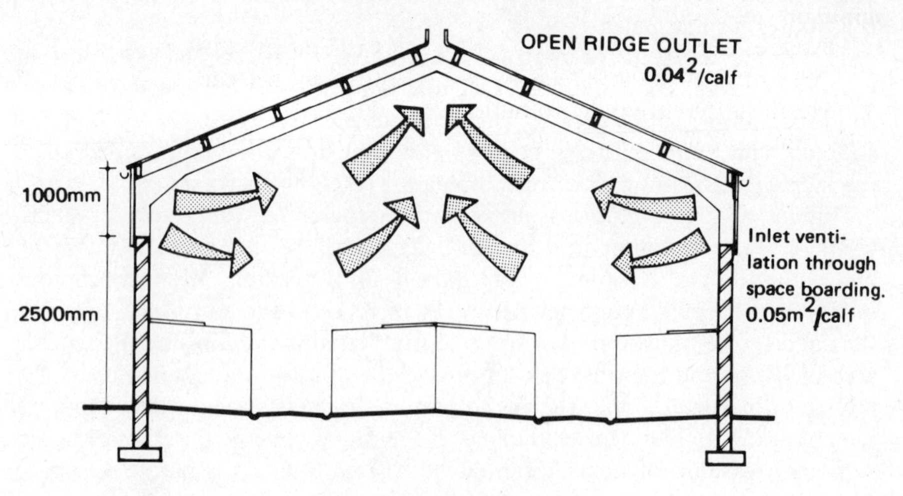

Fig. 1 Ridged roof building

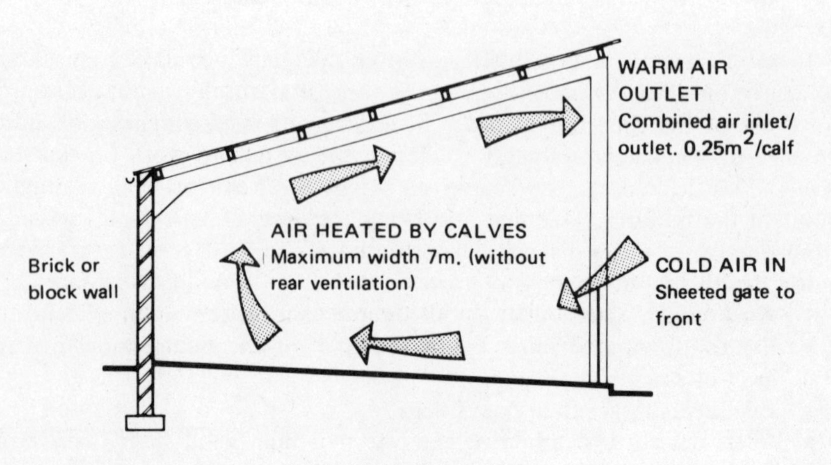

Fig. 2 Mono-pitch building

The precise specifications should take account of individual circumstances but as a general guide for ridged roofbuildings (Fig. 1) the following are the **minimum** required:

Area of outlet at the ridge	=	0·04 m^2 per calf.
Area of inlet	=	0·05 m^2 per calf.
Height difference (inlet to outlet)	=	1·5 m.

In mono-pitched roof buildings (Fig. 2) the air inlet and outlet are combined in a single vertical opening. The size of the opening should be calculated for each individual case but if the opening is 1·5 m high the minimum area required per calf would be 0·15m^2. If the building is over 7 m from front to back air distribution will be improved by creating an adjustable ventilation opening in the rear wall.

1. VENTILATION OUTLETS

Outlets should always be placed at the highest point of the building and in most cases the best solution is to create an open ridge. Although entry of rain and snow is not a major problem with open ridges some moisture does enter and this can be eliminated where necessary by a suitably designed protected open ridge. Alternatives to open ridges are ventilation shafts or slotted roofs.

2. VENTILATION INLETS

Inlets should be designed to provide fresh air uniformly throughout the building without creating draughts. This can best be achieved in most naturally ventilated buildings by having the inlets above calf height uniformly distributed along both sides of the building. A continuous inlet is ideal. Protection against high winds is also required and this is usually achieved by covering the inlet with perforated material such as space boarding or plastic mesh. It is very important to ensure that inlets are sized taking account of the void area of the perforated material.

3. VENTILATION RATE

The effect of different ventilation rates on calf performance at predetermined temperatures was investigated at Drayton and High Mowthorpe EHFs and compared with natural ventilation at ambient temperatures. Ventilation rates varied from 35 m^3 per hour to 170 m^3 per hour per calf place. Diurnal fluctuations in temperature and humidity occurred on all treatments and condensation occurred from five weeks onwards at the low ventilation rate accompanied by a high concentration of ammonia and an unpleasant atmosphere.

The health of the calves was generally satisfactory and growth rates and feed consumption were similar on all treatments. There was no evidence to show that calf performance at any rate of controlled ventilation was better than that obtained with natural ventilation.

Fan Ventilation In most circumstances it should be possible to design an effective natural ventilation system rendering fans unnecessary. In converted buildings however, particularly those attached to other buildings, fans may well be required to ensure good ventilation. Either extraction or pressurised systems can be used as long as the objective of uniform distribution of fresh air at low air speeds is achieved. The most important principle to take into account in the design of fan ventilator systems is that uniform distribution of air can only be achieved by having uniformly placed inlets - whether pressurised or extraction systems are used. Ducting to convey incoming air throughout the length of the building is commonly required to achieve this objective.

The results of the trials carried out at the Experimental Husbandry Farms and elsewhere suggest that where fans are necessary a minimum ventilation rate of 35 m^3/h per calf on arrival increasing to 116 m^3/h per calf at six weeks of age would provide a satisfactory climatic environment.

In summer these rates should be increased to a minimum of 105 m^3/h although rates of up to 205 m^3/h may be beneficial if they can be achieved without causing draughts (D. Michell. Calf Housing Handbook. 1976).

The essential features of management of fan ventilated systems are that the ventilation rate should never fall below 35 m^3/h per calf and since air temperature, as such, is not of primary importance to the healthy calf the use of thermostats to control ventilation rate should be avoided at all costs. The usual result of using this method of control is to provide inadequate ventilation. The best practical method of control is to ensure that the fan throughput cannot be reduced below the minimum required and then use manual manipulation of the controls to ensure that the atmosphere is kept fresh.

Light

Calf accommodation should always make use of natural light. There is no merit in keeping calves in dark buildings. During normal daylight hours all the light required can generally be provided by rooflights. As a rough guide the area of rooflight required in a pitched roof building should be about 10 per cent of the floor area. In addition artificial lighting to give an average level of 50 lux should be available for use by the stockman as required. Whenever stock are housed provision should be made for veterinary inspection at any time and this can be achieved by the use of a portable inspection lamp fitted with a 60 watt bulb. The detailed procedure for determining the most appropriate way of providing the correct level of light in particular situations is given in the Farmelectric Handbook No. 25 entitled *Essentials of farm lighting* and the Farmelectric technical information sheet AGR 5-2 gives guidance on the number, size and types of lighting. See also ADAS Technical Report *Lighting in farm buildings to-day*.

Structural environment

The structural environment includes floor space, floor surfaces, pen divisions and air space. There is a close interaction between structural design and the climatic environment.

FLOOR SPACE

The space provided should allow the calf to perform all its natural bodily functions including grooming. The normal interpretation of this requirement is that the minimum pen width should be equal to the maximum withers height of the calf. Minimum space requirement therefore depends on whether the calves are penned individually or in groups.

Individual pens:	Up to 6 weeks. Area per calf 1·35 m^2 (0·9×1·5 m)
	Up to 8 weeks. Area per calf 1·8 m^2 (1×1·8 m)
Group pens:	Up to 8 weeks. Area per calf 1·1 m^2
	Up to 12 weeks. Area per calf 1·5 m^2
Passages:	Working conditions are greatly improved by providing adequate space between the rows of pens. A clear width of 1·2 m between double rows (Fig. 3) and 1 m between the wall and single row of pens (Fig. 4) is required. If buckets project from the pen front the distance is measured from the outer edge of the bucket.

FLOOR SURFACES

Since calves should always be given bedding the primary considerations are the provision of easily cleaned floors within pens, non slip floors in the passages and adequate floor slopes to give rapid drainage. In most circumstances concrete laid over a damp-proof membrane and finished with a steel float within pens and a wooden float in the passages will provide satisfactory surfaces. Floor slopes within pens should be a minimum of 1 in 20 and a minimum of 1 in 40 in passageways. Floor surfaces should be designed to restrict the wet surface area to a minimum and conduct the liquids outside as rapidly as possible.

PEN DIVISIONS

The choice lies between solid and perforated divisions. Perforated divisions allow social contact and, if the perforations are small enough, prevent vice between calves. Solid divisions reduce the risk of draughts, the spread of enteric diseases and facilitate cleaning. Pen divisions should be not less than 900 mm high and be easily removable for cleansing and disinfection. The front of the pen should always be easily opened and removed. (Fig. 5). Design detail depends on the feeding system.

KENNELLING

Where kennelling is required its sole purpose is to provide the calf with a draught free lying area. To ensure that the ventilation is not unduly restricted only the rear half of the pen should be covered.

FEEDING SPACE

A calf requires a minimum of 350 mm of width for feeding. Individually, penned calves are inevitably provided with more than this but if group

Penning Arrangements

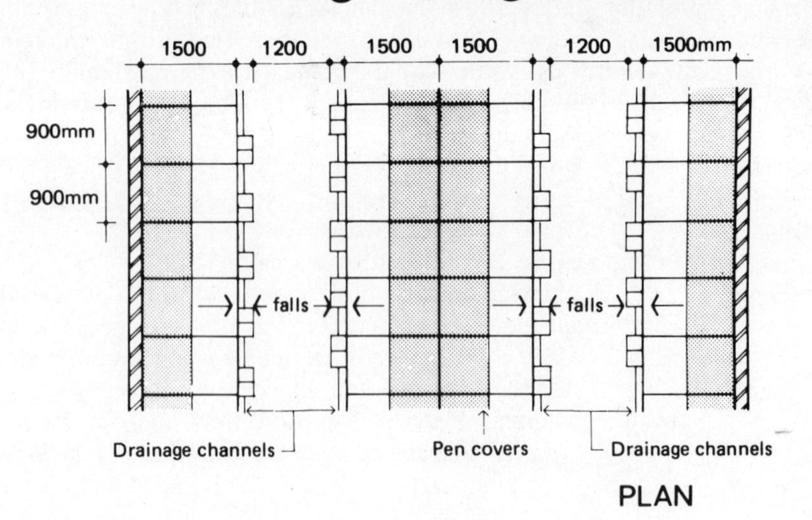

Pen covers 800mm wide

900-
1200mm

falls

falls

SECTION

Fig. 3

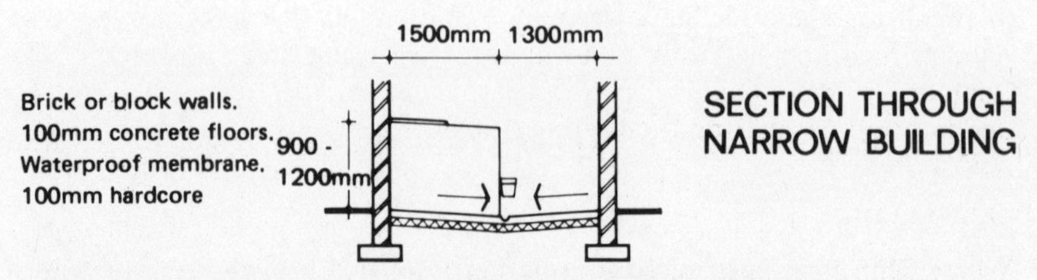

NB. All floors to fall 1:20 to drainage channels in the direction of arrows.

Fig. 4

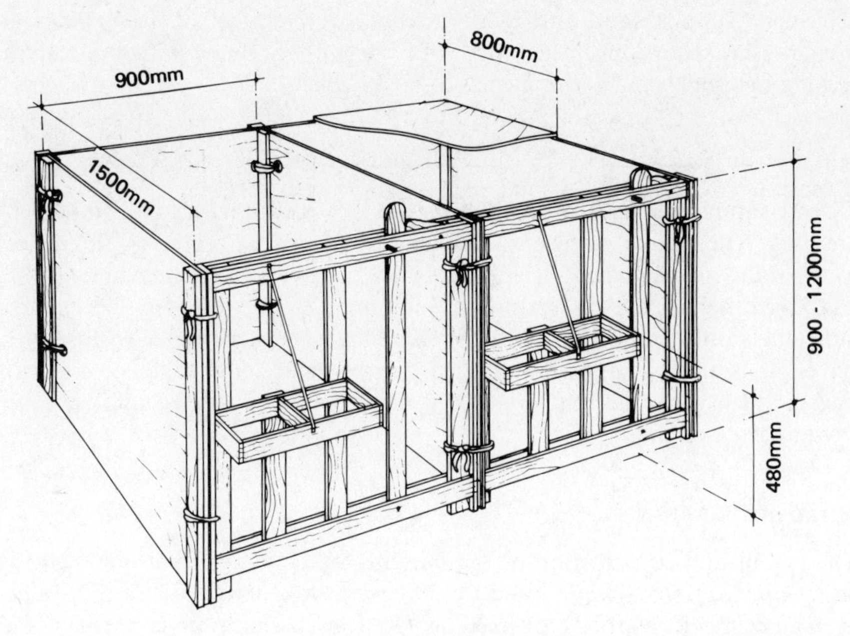

Fig. 5 Individual pens

housed calves are fed simultaneously pen shape may be dictated by feed space requirements.

WATER

Calves should always have a supply of clean, fresh water available. This should be situated close to the drainage channel and where it can be easily cleaned and inspected. Where water bowls are used it is good practice to provide more than one to any group of animals to reduce the risk of supply failure.

AIR SPACE

The lack of sufficient air space has two detrimental consequences to the calf.

1. The provision of adequate ventilation without high air speed is very difficult if not impossible.
2. Exhaled airborne micro-organisms are kept in close proximity to the stock thereby increasing the risk of respiratory infection. A **minimum** of 7 m^3 of air space per calf should be provided, but preferably more.

FEED STORAGE AND PREPARATION

An adequate area should be provided close to but completely separated from the calves. It should be well lit, well ventilated and properly drained. It requires hot and cold water, full washing facilities, secure storage for veterinary supplies and sufficient area for dry storage of feed and feed preparation. The appropriate size will depend on the number of calves, feeding methods and the frequency of feed delivery.

ISOLATION PENS

Disease can be transmitted rapidly between calves even where individual pens are used. The risk of disease spread can be reduced by removing sick calves to separate isolation pens where they can be given supplementary heating and any other necessary treatment. Such pens should be of a size to take individuals only and they should not share a common air space with healthy calves. It is clearly impossible to give unequivocal guidelines on the size of such a facility but accommodation for 3–5 per cent of the total number would be reasonable.

BUILDING DESIGN

The layout and construction of the building must be geared to meet all the basic specifications already listed but must also take into account the system of management, methods of feeding, feed preparation requirements and cleaning out arrangements in addition to site features. Detailed advice on construction can be obtained by contacting LAWS at the nearest divisional office of the Ministry.

Social environment

This is primarily concerned with the social contact provided by the system of management and housing. To some extent there is a conflict between providing the social interaction natural to the stock and the prevention of disease spread.

Young stock should not share a common air space with older cattle. Ideally each age group of calves should be kept in a separate air space and the total number sharing a common air space should not exceed 40–60.

Group size in a single pen can vary between 1 and 20 depending partly on the method of feeding. When calves are penned individually the layout and construction should enable calves to see each other.

Choice of accommodation

Group or individual penning

The first question to be resolved is whether individual or group penning is to be used taking account of feeding method, source of stock and available facilities. The method of feeding has a major influence on the choice because unrestricted liquid feeding is easier to arrange for group housed calves whereas bucket feeding is simplified by individual penning. The source of calves may also influence the choice, since risk of disease spread may be greater with purchased calves which would tend to favour individual pens. If existing accommodation is available for conversion this may also influence the choice of internal layout. Group penning is much more flexible in terms of pen shape and hence will often make the best use of space.

Natural or fan ventilation

The choice here is simple. If the requirements for effective natural ventilation given under 'ventilation' can be met this is the method to use. If these requirements cannot be fulfilled because the building adjoins others on one or both sides or for some other reason, and if this building has to be used then a fan ventilated system is needed. The requirements of the calf remain exactly the same with both systems. The reason for using natural ventilation in preference to fan ventilation is simply that it costs less and is less susceptible to misguided human interference or mechanical failure.

Ridged or mono-pitch roofed buildings

Where new buildings are being considered the choice will lie between ridged and mono-pitch roofs. Mono-pitch buildings have advantages in terms of low cost, ease of access for muck removal, ease of construction and can be easily built in small cells which automatically restricts the number of calves in each air space. They are also very flexible in that they provide excellent follow-on accommodation for calves up to six months of age. They may or may not have a covered passage at the rear but if they do not stockmen will be inclined to speed up routine servicing activities in inclement weather which is not to the advantage of the stock.

If mono-pitch buildings are used for individually penned calves the ventilation opening(s) must be protected by space boarding or similar material to reduce air speeds and avoid the risk of draughts. Siting of mono-pitched buildings is particularly important in that they should face away from the prevailing wind, preferably in a southerly direction and have sufficient clear working area in front for mechanised cleaning out.

Correctly designed ridged roof buildings can provide an excellent, uniform environment with very good working conditions for the stockmen. Within this overall design the choice normally lies between two or four rows of individual pens or a single or double row of group pens. The wider building capable of taking four rows of individual pens makes economical use of space and has particular application on larger units.

Converted buildings

Many different designs of existing buildings can be converted for calf rearing provided the basic principles are observed. The most common problems arising in converting buildings are to do with ventilation, drainage, access, hygiene and isolation.

1. VENTILATION

Satisfactory ventilation can be difficult to achieve by natural means in many situations. In converted cowsheds for example inadequate volume of air per calf (less than 7 m^3) can be a problem when the house is fully stocked. Well designed fan ventilation systems will often be required.

2. DRAINAGE

In existing buildings floor slopes are very often completely wrong, or at the very least inadequate. If liquid is not removed rapidly from the building bedding will get wet, humidity will rise and calves will be predisposed to disease. It is false economy to allow this situation to exist. Floors should be re-laid wherever necessary.

3. ACCESS

Old buildings frequently have very poor access for muck removal. Alterations involved in rectifying this will not always be cost effective, and should therefore be considered carefully.

4. HYGIENE

The surfaces to which calves are exposed should be easily cleaned and disinfected. It is advisable to reduce the risk of disease spread by rendering or other appropriate treatment to enable surfaces to be easily cleaned.

5. ISOLATION

Existing buildings are often in very close proximity to other livestock housing. Steps should be taken to ensure that air-borne micro-organisms are not conveyed by ventilating air, from one group to the next.

Follow-on accommodation

The basic environmental requirements from 6–12 weeks of age are the same as for the younger calf and the ventilation provisions previously stated are sufficient for calves up to 12 weeks of age. Group housing is the normal choice for calves of this size and the additional muck produced increases the importance of ease of access for mechanised cleaning out. The temptation to house calves of this age along with older cattle should be resisted and, for this reason, the open fronted mono-pitch is particularly appropriate.

Husbandry and Disease Control

Calf health depends to a great extent on management. The most important factors are, cleanliness and care at birth (page 2) an adequate intake of colostrum (page 4) careful feeding, particularly during the first few days (page 5) and well ventilated draught free accommodation that can be easily cleaned and disinfected between batches (page 23).

Prevention of ill-health is better than cure and it is generally accepted that the majority of calf diseases arise as a direct result of poor management. As well as being available to treat sick animals, your veterinary surgeon will be able to give appropriate advice on measures to prevent disease.

If the rules of good husbandry outlined in this publication are adhered to sickness and mortality will be considerably reduced. This is particularly true of home-bred calves which have not been exposed to diseases other than those present on the farm where they were born. With bought-in calves the risks are greater but the mortality rate should be no more than 4 per cent.

Cleaning and disinfection

When calf houses are used continuously there is inevitably a build-up of disease organisms resulting in an increased incidence of ill-health and poorer calf performance. This problem can be overcome by efficient cleaning and disinfection and by resting the house between batches. The cleaning process should consist of three distinct operations and should start as soon as the calves and manure have been removed.

1. Remove all dirt from the equipment and structure by pressure hosing and scrubbing with washing soda (1 kg/25 litres).
2. Disinfect the cleaned surfaces by applying an approved disinfectant by brush or spray. Fumigate with a formalin/potassium permanganate mixture only if the building can be sealed effectively.
3. The building should be allowed to dry out thoroughly and rested for as long as possible.

The length of the rest period will depend on the efficiency of disinfection but at least one week should be allowed after the building has dried out before the next batch of calves is brought in.

The ringworm fungus is not destroyed by fumigation and creosote or blow lamp treatment is desirable where this is prevalent.

Advice on Approved disinfectants can be obtained from the Divisional Veterinary Officer and detailed information is given in Leaflet 648 *The cleaning and disinfection of calf houses*.

The use of formalin/potassium permangate mixtures for fumigation purposes involves a fire and health risk and appropriate precautions should be taken.

Bought calves

Care must be taken to avoid the introduction of disease and the source of supply and the type of calf purchased is important in this respect. Ideally calves at least one week old should be purchased direct from the farm of birth in order to minimise stress and reduce the risk of infection. The majority of calves however change hands through the open market and often travel considerable distances to the rearing farm.

All calves should be inspected before purchase and any suspect calves that are dull or show signs of diarrhoea (wet hocks) wet navels, discharges from eyes, nose or mouth, rapid breathing or physical defects should be rejected. On arrival at the farm the calves should be put in clean, dry, well littered pens and allowed to rest. Supplementary heat may be required in cold weather.

Purchased calves often appear hungry on arrival but due to stress factors their ability to digest milk substitutes is impaired and some six hours should be allowed to elapse between arrival and feeding. Where calves have been without food for some time (i.e. after a long journey) feeding a glucose solution (50g/litre of water) as a readily available source of energy is recommended. If any sign of diarrhoea is evident water alone or water plus salt (two teaspoonsful/litre) should be offered, alternatively a proprietary electrolyte solution should be given.

An injection of vitamins A, D and E may be worthwhile for calves born from Feb–April as their dams may have been on vitamin deficient diets in late winter but veterinary advice should be obtained first.

Milk substitutes should be fed according to the makers' instructions.

Sick calves should be isolated in a comfortable environment for treatment.

Calf diseases

Ill-health of calves requires accurate diagnosis and skilled treatment.

Navel and joint-ill

Blood vessels within the umbilical cord pass directly into the abdomen and infection by this route is possible for some hours after birth. Infection results in swollen joints which may cripple the calf, or lead to peritonitis and death.

Cows should be calved in clean conditions and the navel should be treated with an antiseptic such as tincture of iodine as soon after birth as possible.

Scour or diarrhoea

There are many causes of diarrhoea. It may simply be due to faulty nutrition as when, for example, the calf is sucking a cow with too much milk or is given too much milk substitute. It may also occur if milk substitutes are fed too warm or fed at irregular intervals, or if the calf is subjected to too rapid changes in diet. Unless a nutritional scour is corrected it may progress to one in which infectious organisms are involved. Organisms commonly associated with diarrhoea are viruses (rotavirus, coronavirus) bacteria (*E. coli*, Salmonellae, campylobacters) or protozoa (cryptosporidia, coccidia).

Veterinary attention should be sought early in an outbreak of diarrhoea so that the cause may be determined and appropriate action taken. Of particular importance is the prevention of dehydration which is the usual cause of death in calves with diarrhoea. The main signs of dehydration are sunken eyes, a tight skin and coldness to the touch.

Calves with diarrhoea should be isolated where practicable. Milk or milk substitute should be discontinued and only warm water given for one or two feeds. The inclusion of a proprietary balanced electrolyte mixture is desirable in order to combat dehydration. When milk or milk substitute is re-introduced it should be at half strength for one or two feeds. Antibacterial treatment may be given on the advice of a veterinary surgeon.

The incidence of diarrhoea may be greatly reduced by cleanliness and care at birth, ensuring an adequate intake of colostrum, and correct feeding in a comfortable draught free environment.

Salmonellosis

Salmonellosis is a serious bacterial disease which can be spread to man and other animals. Bought-in calves, particularly those that have been exposed in markets or subjected to long journeys and irregular feeding, are the commonest way of introducing infection to clean premises.

Infected calves are dull and reluctant to eat. Scour is a common symptom and faeces may be blood stained. Pneumonia frequently occurs especially in septicemic calves, followed by emaciation and death.

Treatment of infected calves is not always successful and good management and hygiene are particularly important especially during an outbreak. Vaccines against *Salmonella dublin* are available although how they may be used most effectively will depend on individual calf rearing enterprises, and should be discussed with your veterinary surgeon.

Respiratory disorders

A wide variety of infectious agents (i.e. viruses, mycoplasmas and bacteria) occur in the respiratory tract but adverse management/husbandry factors such as overcrowding, poor ventilation and stress may be necessary to precipitate clinical disease.

Respiratory disease is best controlled by paying attention to the environment, the diet and level of stocking. Calves should not be housed in a common air space with older animals. Mouldy hay and straw must not be used and dust in feed should be eliminated as far as possible. On some farms vaccines may be useful.

Calf diphtheria

Calf diphtheria is most commonly found in calves housed in dark dirty conditions where inadequate attention is paid to hygiene, and dirty buckets or other utensils are used.

Affected animals have difficulty in eating and drinking and may dribble from the mouth. Inspection may reveal areas of dead tissue on the tongue, lips or palate and veterinary treatment is necessary or pneumonia may occur.

Lead poisoning

Calves are very susceptible to lead poisoning, the most common sources being lead based paint and roofing felt. Flaking paint from old doors and window frames is particularly dangerous.

Most affected calves go blind, collapse in a fit and die quickly. Some may recover if treated promptly but blindness may persist for some time.

Prevention may be achieved by eliminating lead based painted materials from calf buildings. Old paint which is accessible to animals should be burned off and replaced where necessary with a lead-free preservative.

Lice

Heavy infestations of sucking lice can cause anaemia and appropriate control measures should be taken.

Ringworm

This is a fungal infection which can also affect man. The spores are very resistant to routine disinfection, and creosote or blowlamp treatment of buildings may be necessary. Various treatments for calves are available but oral treatment is expensive.

Calves in poor general health are more susceptible and heavy infections are often a sign of poor management.

Routine operations

When routine operations on farm animals are carried out by stockmen it is important that a high standard of profiency should be attained. Instructions may be provided by your veterinary surgeon or obtained through courses organised by the Agricultural Training Board.

Dehorning

All cattle should be dehorned to reduce the risk of injuries to other stock and to personnel. Dehorning should be done during the first few weeks of life at a sensible time in relation to other management factors so as not to increase stress. It should not be done on the same day that calves are moved or when there has been a sudden change in the weather.

1. CAUTERISATION

The best method of dehorning is to use a cauterising iron. This may be heated by gas or electricity either from the mains or a 12 volt battery. The operation should take place when the horn bud is sufficiently developed for removal but still small enough to fit into the iron bit. This will normally be within the first three weeks of life.

A local anaesthetic must be given before the operation commences. The injection site is the occipital grove which lies behind and slightly above the calf's eye. The hair should then be clipped around each bud so that they are completely exposed. After checking that the anaesthetic has taken effect by applying a clean sharp object to the horn bud, dehorning can commence.

The hot iron is first held at right angles to the head and rotated so as to make a lightbrown ring around the base of the bud. It is then held at 45° to the head and the handle of the iron moved in a circle so that the bit travels around the base of the bud and undercuts it. The bud is gouged out by digging the rim of the bit under the bud. Finally if there is any bleeding the wound should be lightly cauterised. A high standard of cleanliness must be maintained throughout the operation.

Other methods of dehorning such as the use of caustic potash or flexible collodion are less reliable and the chemical must be applied before the calf is one week old, as required by the Protection of Animals (Anaesthetics) Act 1954 and 1964.

2. CAUSTIC POTASH

The hair should be clipped from the horn bud and a ring of petroleum jelly should be applied around the outer rim of the horn bud to prevent the spread of caustic. The centre of the horn bud should then be rubbed with a moistened caustic stick for 15 seconds and the application repeated after an interval of five minutes until a slight indentation has been made in the centre of the horn or a trace of blood appears.

It is essential that the potash stick is not moistened too much or the corrosive solution will spread to other areas, possibly the eyes, with harmful results. Appropriate precautions must also be taken by the operator to prevent damage to the fingers.

This method is not recommended for calves suckling cows.

3. FLEXIBLE COLLODION

The hair should be clipped from the horn bud and the area rubbed with a cloth moistened with methylated spirits or other defatting solvent. A small amount of flexible collodion should be rubbed on the horn bud then a second application should be painted on and allowed to dry. The horn bud should be examined on succeeding days and a further application of collodion made if the film is broken.

The first application of collodion should be made before the calf is two days old.

Castration

Unless intended for bull beef production, male calves should be castrated within the first two months of life. (Above this age only a veterinary surgeon or a veterinary practitioner may castrate a bull (Veterinary Surgeons Act 1966).)

The three main methods are the rubber ring, the burdizzo and the knife. Whatever the method a high standard of cleanliness must be maintained.

1. THE RUBBER RING

Rubber rings must, by law, be applied within the first week of life. With the help of a special instrument (elastrator), the ring is passed over the testicles and released above them onto the upper part of the scrotum, taking care that no teat is trapped. The operation is easy to undertake, but causes considerable stress and sepsis is fairly common.

2. THE BURDIZZO CASTRATOR

The testicles are pulled to the bottom of the scrotal sac and one of the spermatic cords is then held against the side of the sac, placed in the jaws of the castrator and crushed in two different places. The second crush must always be made below the first. The operation is then repeated on the other cord. Crush marks must never join across the scrotum and a gap of about 12 mm always being left by varying the position of the castrator. This method has the advantage of not causing an open wound which may lead to infection.

3. THE KNIFE

The knife must be sterile and sharp. The scrotum should be cleaned and disinfected by swabbing with antiseptic solution. the testicles should be grasped and moved to the bottom of the scrotum. A sufficiently long incision should then be made along the bottom of the scrotum in a direction away from the operator so as to expose both testicles. Each testicle should then be squeezed out in turn and the cord and vessels twisted before breaking them by pulling. They should not be cut with the scalpel.

With calves that are over five weeks old the cord will have to be separated and cut but the vessels should be twisted and broken as above. The wound should be dusted with antiseptic powder. The method has the advantages that stress is limited to a short period and that it is foolproof, its disadvantage is that the wound may become septic, for this reason clean bedding is an essential. Fly repellant cream should be applied when necessary.

Removal of surplus teats

Spare teats on heifers intended for breeding detract from their appearance and may interfere with milking machine performance. They should be cut off with a pair of sharp, sterile, bluntnosed, curved scissors when the calves are about one month old. The site must be disinfected before and after the operation. Laymen are not permitted to remove supernumerary teats on calves over three months old.

Identification

Under Article 16 of the Tuberculosis Order, 1964 (S.I. 1964 No. 1151) all cattle (with the exception of calves aged 14 days or under, which are moved directly to a slaughterhouse) are required to be marked or otherwise identifiable in accordance with the National Herd Marking Scheme (or in the case of pedigree animals, in accordance with the rules of a breed society).

Under this system, the Divisional Veterinary Officer allocates each herd with a herd mark. This is a letter-number combination, allocated on a county basis, and must be inserted in the right ear of each animal in the herd.

Animals in a herd allocated, say, the herd mark BA297 would be marked BA297-1, BA297-2, BA297-3, etc. If the right ear of an animal is so deformed or otherwise unsuitable that the mark cannot be placed in it, it is permissible to place the mark in the left ear.

The animal must be marked by means of a tattoo or an approved ear tag bearing the herd mark and individual animal mark. The Divisional Veterinary Officer will advise on the manufacturers and types of approved ear tags.

The animal mark for calves may consist of a letter representing the year of birth and a serial number denoting the order of birth in that year. Under this system calves would be marked for example, BA297-A1, BA297-A2, etc. in one year and BA297-B1, BA297-B2 in the next year. The letters I, O, R and U should not be used for denoting the year of birth.

A high standard of hygiene must be maintained when undertaking ear tagging or tattooing so as to prevent sepsis. Tags should be inserted so as to allow sufficient room for the ear to grow.

Further information on methods of identification can be obtained in MAFF Leaflet 600 *Cow identification* and in the Agricultural Training Board Leaflets MPB I.C.1 and 2.

Appendix I

Basic requirements

1. Temperature – No specific requirement for healthy calves. Supplementary heat may be required for sick, stressed or starved calves.
2. Relative humidity – Similar to outside relative humidity.
3. Ventilation – Climatic housing.

   Pitched roof: Eaves inlet $0{\cdot}05$ m^2, per calf ($0{\cdot}5$ ft^2)
   Ridge outlet $0{\cdot}04$ m^2 per calf ($0{\cdot}4$ ft^2)
   Height difference (inlet to outlet) 1·5–2·5 m (5–8 ft)

   Mono-pitch: Narrow span: single opening size calculated for individual situation and depends on the dimensions of the building and the number of calves housed. Not less than $0{\cdot}15$ m^2 per calf. Additional wide span:– perforated opening in the rear wall of mono-pitch buildings more than 7 m wide improves air distribution.

   Fan ventilation: Extraction and pressurised systems: Only required in exceptional circumstances. Minimum ventilation capacity 35–105 m^3/h per calf (24–73 ft^3/min). Use manual control.
4. Air speed at calf level – Maximum under cold conditions 0·25 m/s (50 ft/min)
   Use kennelling to prevent down draughts. Inlets above calf level.
5. Air space (cubic capacity) – Minimum of 7 m^3 per calf (250 ft^3)
6. Floor space – Individual pens: Up to 6 wks $1{\cdot}5\times0{\cdot}9$ m (5×3 ft)
   Up to 8 wks $1{\cdot}8\times1{\cdot}00$ m ($6\times3{\cdot}28$ ft)

   Groups: Up to 8 wks $1{\cdot}1$ m^2 (11.84 ft^2)
   Up to 12 weeks $1{\cdot}5$ m^2 ($16{\cdot}1$ ft^2)
7. Feeding space – minimum 0·35 m per calf (1·15 ft)
8. Drainage – floor slope at least 1 in 20 under calves.
   floor slope at least 1 in 40 in drainage channels.

Appendix II

Production targets

Birth weights

Friesian calves from cows	=	42 kg
Friesian calves from heifers	=	37 kg
Medium beef breed × Friesian	=	37 kg
Large beef breed × Friesian	=	45 kg
Av. buying-in weight Friesian bulls	=	40–45 kg

12 week weights

Friesian heifer calf	=	85–95 kg
Friesian bull calf	=	100–110 kg
Beef × Friesian heifer calf	=	85–100 kg
Beef × Friesian bull calf	=	100–110 kg

Feed use

Milk substitute	=	12–14 kg
Early weaner mix	=	130 kg
Hay	=	9 kg

Appendix III

Budgeting Pro Forma

(Rearing to 12 weeks)

OUTPUT		
Sale price		____
Less		
Calf cost		____
Mortality cost		____
Net output		____
VARIABLE COSTS		
Milk Powder	(____kg @ ____)	____
Concentrates	(____kg @ ____)	____
	(____kg @ ____)	____
Hay and other feed	(____kg @ ____)	____
Straw	(____kg @ ____)	____
Vet and medicine costs		____
Miscellaneous costs		____
Market and transport costs		____
Total variable costs		____
GROSS MARGIN PER HEAD		____

When budgeting a new unit, the fixed costs for the unit, i.e.: building charges, interest on capital, labour use, power, heating and other changes must be deducted from the gross margin for the enterprise.

Appendix IV

MAFF Publications

Calf housing L 759;
Cleansing and disinfection of calf houses; L 645
Codes of recommendations for the welfare of livestock. Cattle. L 701
Causes and prevention of losses in calf rearing L 517
Gleadthorpe EHF. *Beef Bulletin*
Liscombe EHF *Beef Bulletin No 1 Calf Rearing*
Cow identification L 600

Appendix V

Other publications

Calf Housing Handbook, by C D Mitchell. Available from Scottish Farm Building Investigation Unit, Craibstone, Bucksburn, Aberdeen, AB2 9TR

The Calf, by J H B Roy, two volumes (Iliffe, London).

Webster, A J F, 1981. *Optimal Housing Criteria for Ruminants.* **In**: Clark, J A, (Ed.) *Environmental Aspects of Housing for Animal Protection.* (Butterworths).

Agricultural Training Board Leaflets. MPB 1.A. 1–4; MPB, 1.B. 1–10; MPB 1.C. 1–9. Available from Agricultural Training Board, 32 Beckenham Road, Beckenham, Kent, BR3 4PB

Appendix VI

ADAS Experimental Husbandry Farms where calf rearing work is being carried on with dairy bred calves

Boxworth EHF, Boxworth, Cambs. CB3 8NN
Bridgets EHF, Martyr Worthy, Winchester, Hants. SO21 1AP
Drayton EHF, Alcester Road, Stratford-on-Avon, Warwicks CV37 9RG
Gleadthorpe EHF, Meden Vale, Mansfield, Notts. NG20 9PF
High Mowthorpe EHF, Duggleby, Malton, Yorks. YO17 8BW
Liscombe EHF, Dulverton, Somerset TA22 9PZ
Rosemaund EHF, Preston Wynne, Hereford HR1 3PG
Trawsgoed EHF, Trawsgoed, Aberystwyth, Dyfed SY23 4HT

Appendix VII

Conversion Tables - Metric to Imperial

Tables for rapid conversion of figures used in this publication

1 metre (m)	=	1·094 yd
1 millimetre (mm)	=	0·0394 in
1 square metre (m^2)	=	10·76 sq ft
1 cubic metre (m^3)	=	35·3 cu ft
1 litre	=	0·22 gallon
	=	1·76 pints
1 millilitre (ml)	=	0·035 fluid oz
1 litre of milk weighs	=	1·03 kg
100 g of milk substitute per litre	=	1 lb per gallon
1 tonne	=	0·98 ton
1 kilogram (kg)	=	2.2 lb
1 gram (g)	=	0·035 oz
1 bar	=	14·5 pound force per sq in (lb f/ in^2)
(°C × 1·8) + 32	=	°F
4·7 kg milk	=	10·32 lb = 1 gallon
4·5 litres	=	1 gallon

Printed in the UK for HMSO
Dd737452 C30 6/84 (3062)